THE INVENTION OF THE WATCH PHONE

TIME WAITS FOR NO ONE

DOMINIQUE ROSS & SIVVONG BE

VICIOUS PRODUCTION
Published by Dominique Ross & SivVong Be
London * Beijing * Sydney * New York * Atlanta

Copyright © 2007 by Dominique Ross & Sivvong Be

All rights reserved
Including the right of reproduction
in whole or in part in any form.

Vicious Production, 2007
Published by Dominique Ross & Sivvong Be
Po Box 721102
Dallas, Texas 75372

Designed by Nate McLaughlin

Photography by Kenyon Nahvipour

Manufactured in the United States of America

A cutting-edge and innovative new product, the Watch Phone, has been developed by Sivvong Be and Dominique Ross. The invention's unique design provides users with unprecedented functionality and convenience in regards to the use of a common communications device.

The Watch Phone is cost-effective and will also eliminate some common hazards associated with the attempted use of certain devices while one is engaged in a primary activity. With its extensive collection of redeeming qualities, it should have no trouble finding success through outlets and catalogs selling communications devices.

This original idea is now being made available for licensing to manufacturers interested in new product development, especially in the communications devices industry. Be and Ross are hoping to have the Watch Phone in full production and available to the public within the very near future.

Additional information about Watch Phone can be obtained by contacting the Publicity/Press Department of Invention Technologies, Inc. at (800) 940-9020 ext. 2285 or at products@invent-tech.com. Invention Technologies, Inc. is a Coral Gables, Florida-based firm that is handling the publicity and public relations for Watch Phone.

NOTE: This information is offered as a human-interest story about the inventor and conception of the idea. Neither this release nor any publication of it constitutes disclosure of functional or structural details of the invention.

General Information

Report Contents 4
Invention Industry Overview 5

Product Information

Inventor Profile 7
Product History 7
Product Description and Usage 8
Graphic Illustration 10
Product Benefits and Advantages 11
General Field-of-Art Search of U.S. Patent Database* 12

Industry Information

Overview of U.S. Economy 13
Industry Outlook 21
Trade, Industry and News Media Reports 33

Production

Manufacturing and Production Considerations 45
Cost and Price Estimates 46
Packaging 50
Advertising and Sales Channels 52
Association Database 56

Conclusions

In the future hand-held phone will be a safety hazard, obsolete, and outlaw because it will be illegal to talk and drive with a phone. The Watch Phone will be the new face of the future in technology in mobile phones. Developed by Sivvong Be and Dominique Ross, the invention of the Watch Phone is a unique concept based upon real actual daily incidence. The design is to act as a phone and a watch to create not only safety but functionality, practicality, style, class, and data.

The Watch Phone is cost-effective and eliminates common hazards associated with the attempted use of mobile devices when one is engaged in a primary activity such as driving an automobile. The Watch Phone will become a primary source for paying bills and buying groceries. The Watch Phone is good for fitness endeavors from jogging to playing tennis. The Watch Phone acts as a tracking device for children by accessing via internet. The use of fiber option Bluetooth headset gives the Watch Phone the capability to be use as a remote for television set and hearing device. The headset acts as hearing aide and with any 2008 automobile sound system. Professionals such as such as fireman, doctor, or the military will find the invention useful for its camera or cam recorder feature which allow interactions via chat or live camera (require a computer with the watch phone compatibility).

We would like to give compliments and tributes to all inventors who have diligently created products similar to the watch phone. All the data is a breakdown of the formula to the actual Watch Phone showing benefits, costs, and comparisons to actual watch

phone. We welcome you to read and learn more about this fantastic and innovative technology.

The Dick Tracy Show

M500 WATCH PHONE

China M800 Phone Watch

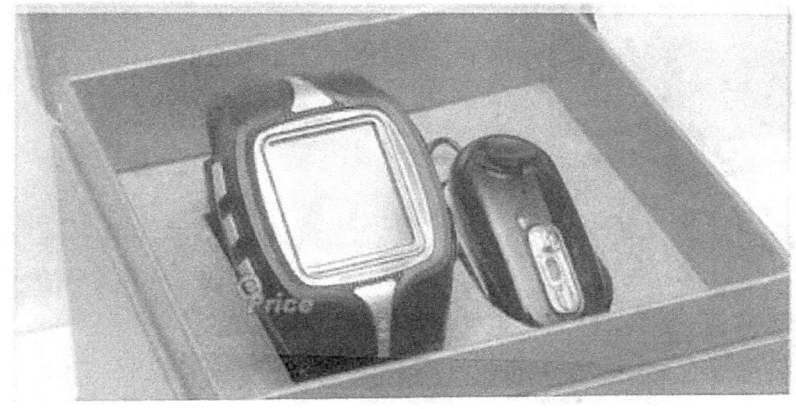

GSM-F88

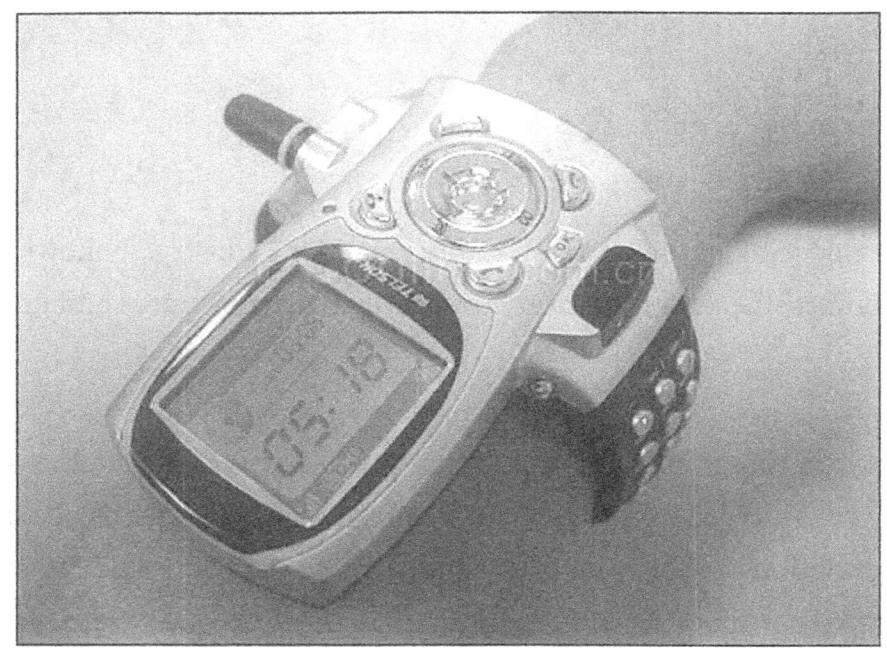

TELSON

Report Contents

This report begins with a brief overview of the invention industry. Following are details about the inventors, the invention, and how it will affect consumers and its related industry.

In order to create a comprehensive representation of the product and its market, sections describing the evolution of the product, its benefits and uses, a technical overview and possible design options are provided. Manufacturing and production details, packaging considerations, cost estimates, sales and advertising options, and related articles from various media are also examined. Furthermore, related industry data, including a U.S. market overview and an industry market overview, provides a global picture of the current economic environment. In covering these topics the report does not propose to be all-inclusive of the factors involved in preparing an invention for potential development.

The report also contains an introductory graphic illustration prepared by experienced draftsmen. In addition, a preliminary General Field-of-Art Search of the United States Patent Database has been conducted by a Patent Attorneys, and similar or relevant patents have been included.

Invention Industry Overview

Invention is the creative act that results in a process, device or technique which represents a unique or innovative application of existing technology or advances the boundaries of technology through study and development. It is the application of the idea or invention that is fundamental. The inventor not only conceives the idea but also puts it into practice.

Progress, as we think of it, is largely due to inventions. Every new method, machine, device or system invented adds to our collective wealth of knowledge. Invention began so long ago in our history that we cannot trace its beginnings, but the hallmarks of the human effort to invent are easy to see. The simple yet profound invention of the wheel, for instance, still reverberates in our everyday lives. From it came everything from rollerblades to racecars, steering wheels,

water wheels, potters' wheels and flywheels for engines, to name just a few of its many applications.

Inventions are a result of human beings striving to improve their quality of life. Every invention is a building block to achieving greater knowledge of our world and solving difficulties in our day-to-day lives. From primitive tools to modern day computers, each invention has changed our social structure and way of life. Inventions have stimulated our minds, and given us comfort and convenience, becoming a contribution to our society and its progress.

The English Statute of Patents and Monopolies in 1623 gave the right of patent protection to the first inventor of a new technique or device. This set the stage for the Constitution of the United States (in Article I, section 8, clause 8) to include empowerment of Congress "to promote the progress of science and useful arts by securing for limited times to authors and inventors the exclusive right to their respective writings and discoveries." This helped create the current atmosphere that encourages and supports the inventive process.

There are two theories on how the inventive process arises. The deterministic theory of invention contends that when technical, cultural and economic conditions are right, one person or another will make an invention; who actually does it is simply historical accident.

| Watch Phone | Invention **Industry** Overview |

Many instances of concurrent and independent invention support this idea. The Bessemer-Kelly process of steelmaking (1857) and the Hall-Heroult process for reducing aluminum (1886) were each conceived independently by two different inventors in a short period of time. Kentucky iron manufacturer, William Kelly, and British inventor, Sir Henry Bessemer, were both inspired and receive credit for a process that enabled the inexpensive steel production which supplied the emerging Industrial Age. Similarly, American, Charles Hall and Frenchman, Paul Heroult both arrived at the critical electrolytic process that still remains the only method used for smelting aluminum in the industry. As these examples illustrate, independent inventors in distant countries realized each process almost simultaneously using the available technologies.

Equally important is the individualistic theory of invention, which points to the many occurrences of sudden insight or creative genius that leap beyond the given. Oliver Evans produced a working steam road vehicle in 1805, well ahead of his time's capacity to produce or apply this technology; yet, by doing so, he helped pave the way for the thriving modern automotive industry.

In either theory, the operative factor is still the individual inventor. Without the drive and imagination of each inventor, we could never move forward.

Product Description and Usage

Following is a physical description of the **Watch Phone**, designed to properly characterize the working functions of the product as well as define its physical appearance. Based on the inventors' description and any pertinent additional information obtained, typical use of the product, its basic design and production materials are considered. Overall production considerations are also discussed.

The **Watch Phone** (see detailed drawing) is an innovative new electronic device, worn on the wrist, which functions as a phone, GPS tracking device, audio/video/internet accessory, and more! This cleverly designed product combines the convenience of a wristwatch with the technology of a computer to provide a single device with many beneficial features which enhance utility, enjoyment, and safety alike. It will likely be similar in appearance to a molded plastic wristwatch which communicates with a lightweight, waterproof wireless earpiece that fits comfortably behind the ear. Features can include a fully functional phone with keypad, calculator, GPS tracking device (especially beneficial for children), audio/video/internet capability, mic and speaker voice recognition, hearing aid, and digital camera. The user should be able to access music, games, radio, television, and internet with this innovative wireless product. Industry standard materials (molded plastic, nylon, rubber) and electronic components can be utilized in construction. A variety of popular colors and styles can be offered. All applicable safety standards should be met.

As described, the design parameters for the **Watch Phone** should likely be open to current production processes and tooling. Any materials specified by the inventors are commonly available on the commercial level. Any concerns that may arise during the development of the **Watch Phone** would likely be amenable to resolution through industry-standard product testing and refinement. Whether it would ultimately be feasible to do so from an economic standpoint is a question to be decided by the potential manufacturer and/or licensee. Sometimes, this decision takes place after additional studies are conducted that go beyond the scope of this Invention Research Portfolio.

In order to realize the potential and applications of an innovation, attention should always be given to cross-utilization and further modification of a new product. Straightforward modifications that would significantly expand the utility or usability of the invention are of particular interest. A modification could accomplish this by either further extending the original function of the product or by actually adapting the product to be used separately. Consideration of any modifications should address both materials and manufacturing processes, especially if the modifications require any significant changes to the initial material and manufacturing parameters. Not all products lend themselves easily to either stylistic or design modification.

Given the nature of the design of the **Watch Phone**, any major technical modifications needed to bring the concept to production readiness would be made based on a determination made by the manufacturer and any such observations are not included in this portfolio.

Given the previous description of how the **Watch Phone** would function and taking into consideration the questions facing further development of the proposed product, a production version of this item could potentially be developed to perform generally as the inventors state. Detailed diagrams and technical specifications remain to be drawn. Nonetheless, we believe that we have a clear idea of the inventors' intent and goals in submitting this design. Please note the graphic illustration following this section for a more complete representation of the invention.

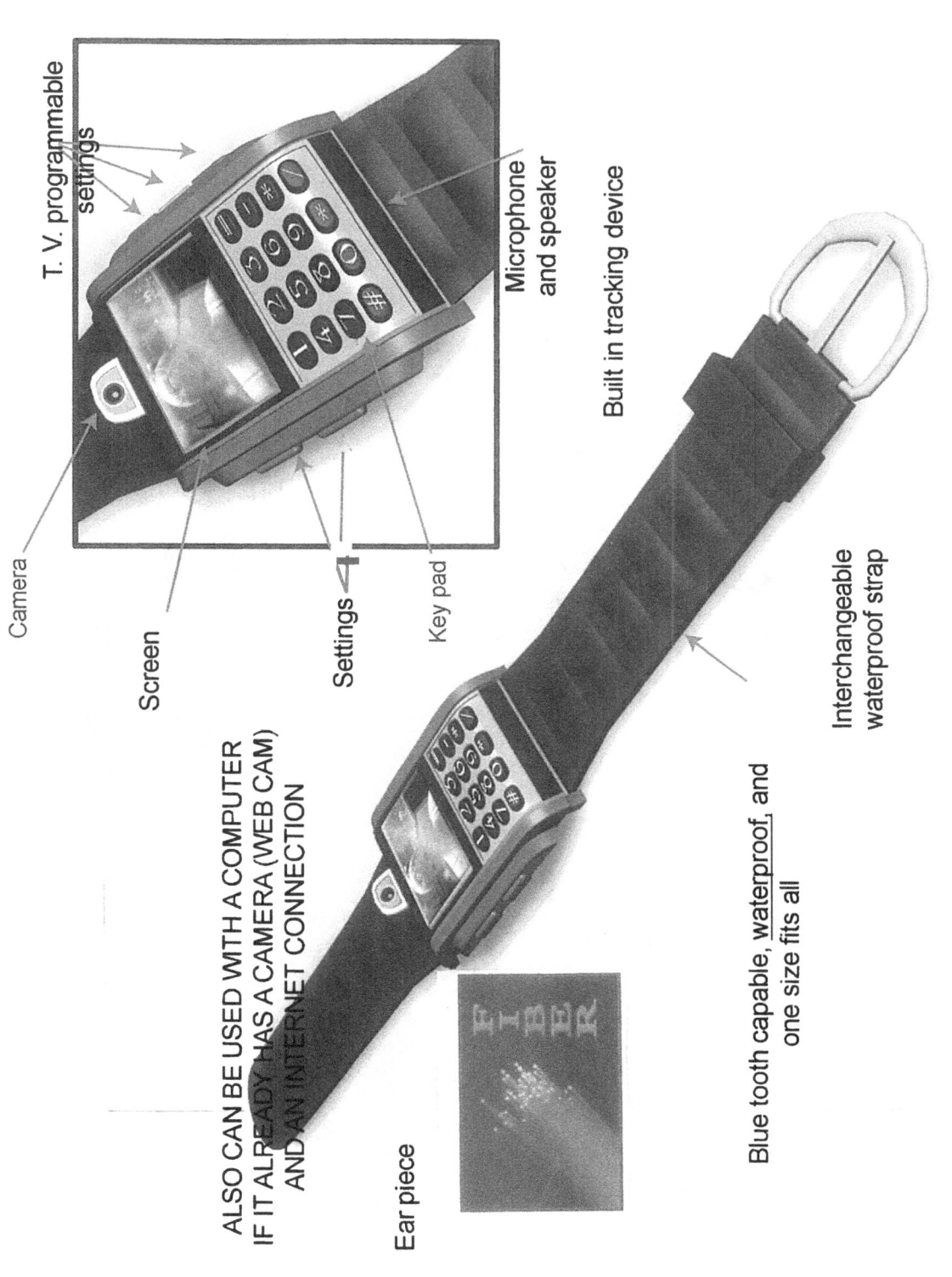

T. V. programmable settings

Microphone and speaker

Built in tracking device

Camera

Screen

Settings

Key pad

Interchangeable waterproof strap

ALSO CAN BE USED WITH A COMPUTER IF IT ALREADY HAS A CAMERA (WEB CAM) AND AN INTERNET CONNECTION

Ear piece

Blue tooth capable, waterproof, and one size fits all

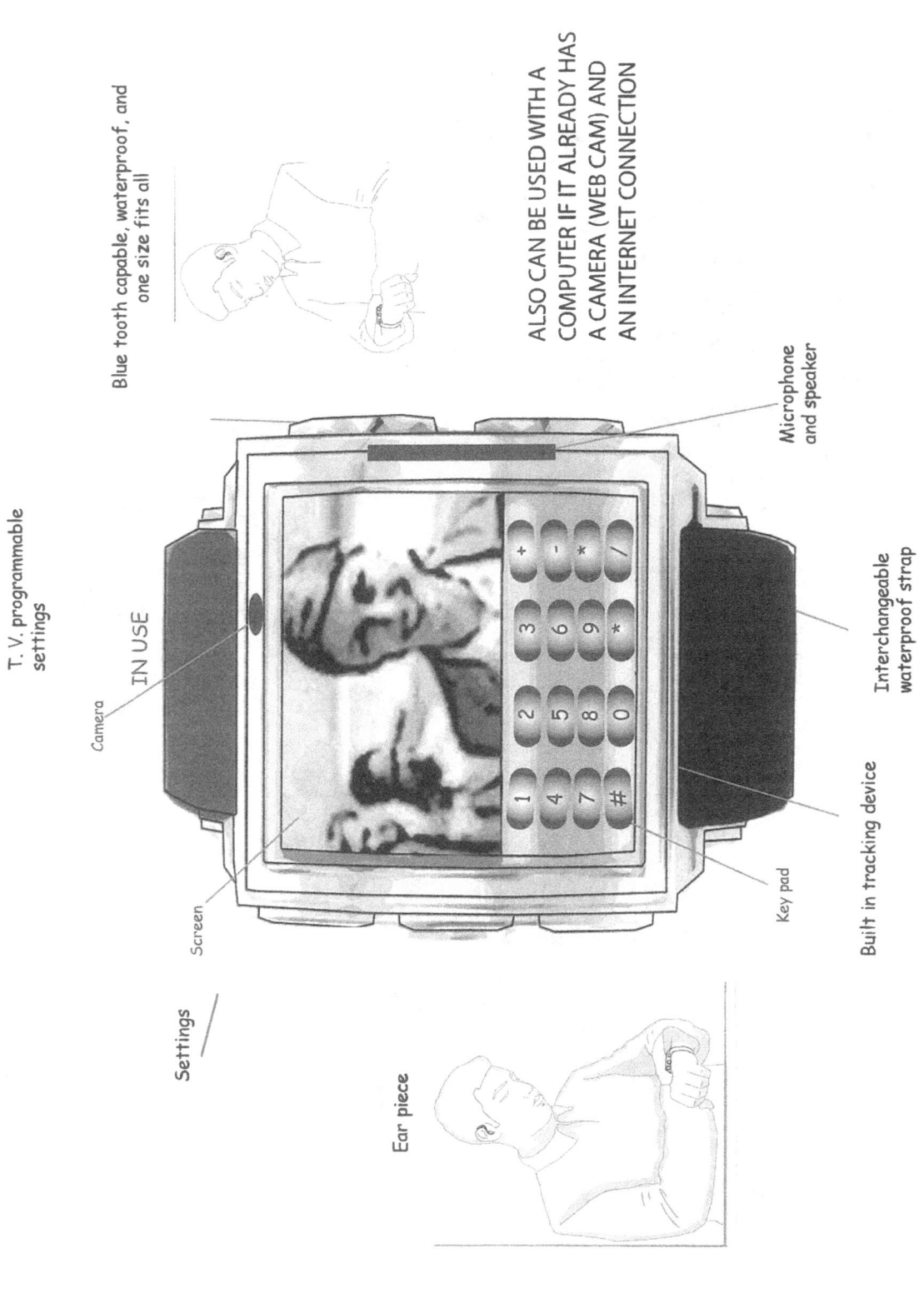

Graphic Illustration

Sketches and visual representations are among the best ways to explain the design and functional advantages of the product. A graphic illustration is able to draw attention to the "visual claims" of the idea. From a layperson's point of view, it also helps establish the unique features of the invention compared to any known similar products.

The illustration may also prove to be useful in the future when presenting the invention to industry. Additionally, this graphic illustration may be forwarded to a patent attorney to assist in the preparation of possible future patent filings.

Please keep in mind that the graphic illustration is merely a representation of the idea. It is not meant to portray the final design specifications or rendering of the invention. Due to the multitude of factors that must be considered prior to launching a new product, manufacturers and their design team will ultimately determine final product design.